Developing Printed Circuit Assemblies

From Specifications to Mass Production

by Elaine Rhodes
and Paul A. Spalitta

First Lulu edition July 12, 2005
Released to general distribution May 20, 2008

ISBN 978-1-4357-1876-0

Photographs on cover:

NEC Monitor: http://www.necmitsubishi.com/images/G/H/LCD1765_left_big.jpg

Canon Digital Camera: http://consumer.usa.canon.com/app/images/eos/rebel_ti_586x225.jpg

DVD Player: http://travelproducts.com/store/enter.htm

This book was produced on a Windows PC using Microsoft Word 2003 and Adobe Acrobat.
The covers were prepared using Adobe Illustrator and Adobe Photoshop.
The text is set in Goudy Old Style, **Lucinda Sans**, and Tahoma.
Century Schoolbook, Tahoma, and **Georgia** are used on the covers.

Contents

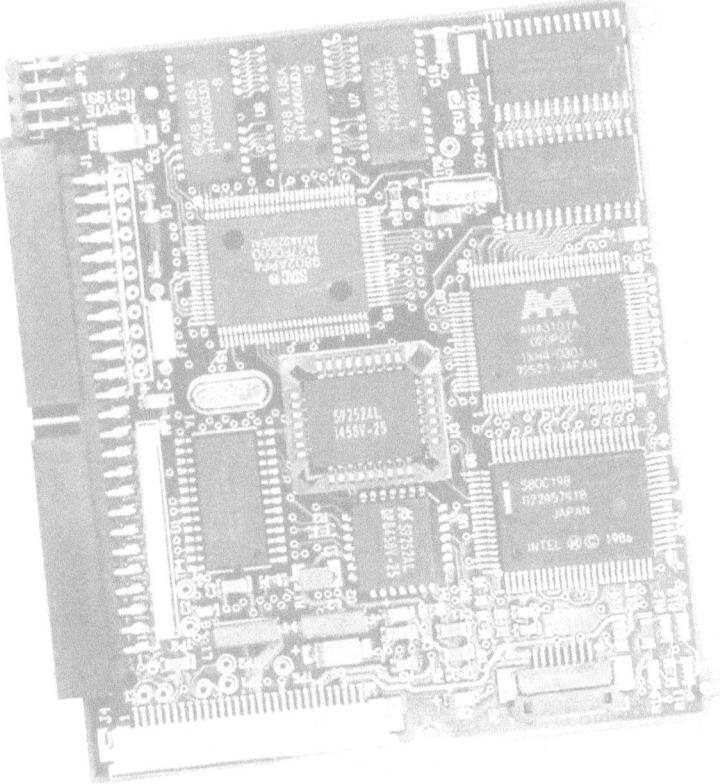

Getting Started

Printed circuit assemblies are the brains inside your
computer, DVD player, digital camera and just about
every other electronic device you own. A printed
circuit assembly (PCA) is a printed circuit board
(PCB) with attached components such as processor chips,
memory, resistors, capacitors, transistors, switches and heat
sinks. The PCB is typically a flat, rigid, fiberglass board with
electronic circuit traces etched in copper on the top, bottom,
and internal surfaces. The PCB provides the structure for the
PCA and interconnects the other components, allowing them
to communicate. PCAs come in many shapes, from rectangular
to intricately carved, and sizes, from dime-sized to the size of a
large pizza box. A sample PCA is shown on the next page and
some of its components are identified.

Sample Printed Circuit Assembly (actual size)

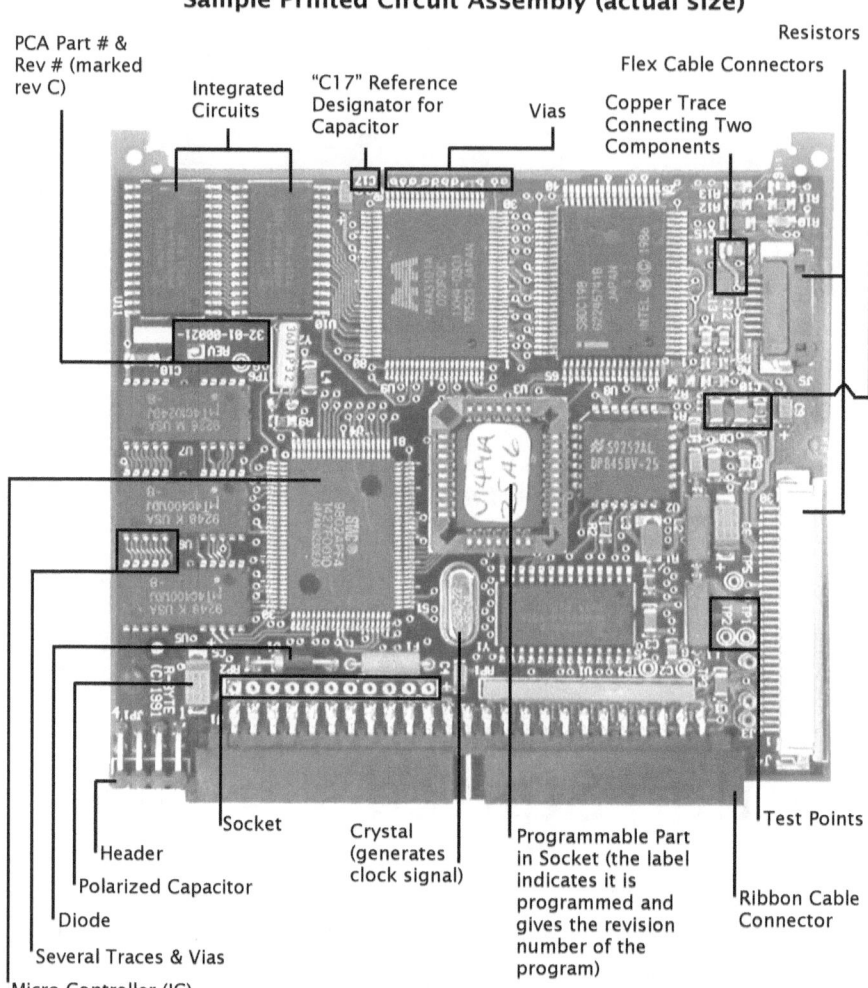

PCA Part # & Rev # (marked rev C)

Integrated Circuits

"C17" Reference Designator for Capacitor

Vias

Resistors

Flex Cable Connectors

Copper Trace Connecting Two Components

Socket

Header

Polarized Capacitor

Diode

Several Traces & Vias

Micro Controller (IC)

Crystal (generates clock signal)

Programmable Part in Socket (the label indicates it is programmed and gives the revision number of the program)

Test Points

Ribbon Cable Connector

Developing a PCA is a time-consuming and complex process. If you are an engineer, this guide can help you learn to approach the development process systematically, from defining specifications to designing the PCA and ultimately handing it off to manufacturing. If you are a program manager, this guide can help you to work confidently with your engineering and marketing teams to save time and money and create high quality PCAs.

2

This guide explains the activities involved in developing PCAs, the purposes of the activities, and how the activities relate to each other. The following topics are outside the scope of the guide:

- How to design electronic circuitry
- How to schedule activities
- How to select vendors for parts and services

Audience

This guide describes the PCA development process from the perspective of the electrical engineer who works as the lead PCA designer. The primary audience for this guide includes engineers who:

- Graduated recently from college with a Bachelors degree or a Masters degree in electrical engineering or a related field
- Worked only in start-ups or small companies and want to improve their companies' PCA development processes
- Worked with PCA development practices that are not up to date with current industry best practices
- Designed PCAs within a different environment such as a research laboratory, academia, or as a hobby
- Designed semiconductors or firmware but not PCAs

This guide is particularly useful if you are a lead PCA design engineer responsible for both designing the PCA and managing the development process, a dual role common in today's cost-driven business environment.

Secondary audiences for this guide include program managers, executives, and others who need a clear understanding of the PCA development process.

Overview of Development Process

The PCA development process consists of the following steps:

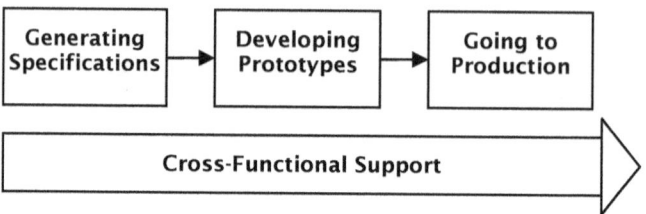

Most PCA development projects are completed in fifteen months to two years, with a year and a half being typical. Only the simplest PCA can go from the specification stage to pilot production in less than nine months. If the specifications are not agreed upon in two or three months, or if they are significantly changed later in the development cycle, the schedule may be delayed.

Most PCA development activities are not strictly sequential; you can overlap some of the tasks to complete the project more quickly.

Tools Needed

The chart on the following page lists the tools used to develop PCAs and manage the development process.

Once you have assembled the tools you need, begin developing your PCA by generating specifications describing all of the PCA's requirements, as discussed in the next section.

Tools Used to Develop PCAs

Tool	Purpose	Example	Note
Word processor	Prepare specifications and text documents	Microsoft Word	
Spreadsheet	Prepare bill of materials and table-based or calculation-intensive documents	Microsoft Excel	
Scheduling program	Generate project schedules	Microsoft Project	You can also use a spreadsheet or a word processor.
Schematic entry program	Draw schematic diagrams	OrCAD, Viewlogic	For convenience, you may use the same brand tools for both schematic entry and PCB layout. This convenience may be outweighed by one tool being superior to another for a particular task. Moving between one vendor's schematic entry tool and another vendor's PCB layout tool is easy because file formats are standardized and conversion utilities work well.
Printed circuit board layout program	Design printed circuit boards	PADS, Mentor	
Mechanical design program	Used by mechanical engineers to design mechanical parts	AutoCAD for 2-D design, HP Solid Designer for 3-D modeling	You may work with files generated by these programs to access mechanical design information that bears on the PCA design. In some cases you may use these tools yourself.
Computer with Internet connection	Access component and technical information and share documents	Varies by organization	
Lab tools: oscilloscopes, DVMs, soldering irons, microscopes, power supplies, jigs, and fixtures	Bring-up, debug, test, rework (fix and rewire by hand), and repair PCAs in the lab	Varies by organization	Due to their sophisticated nature, further description of these tools is beyond the scope of this guide – consult with each tool's manufacturer for more information about that tool.

Note: Always use your organization's standard software, if a standard exists. If you need to use a different program, be certain its file format interchanges with the organization's standard so you can share documents.

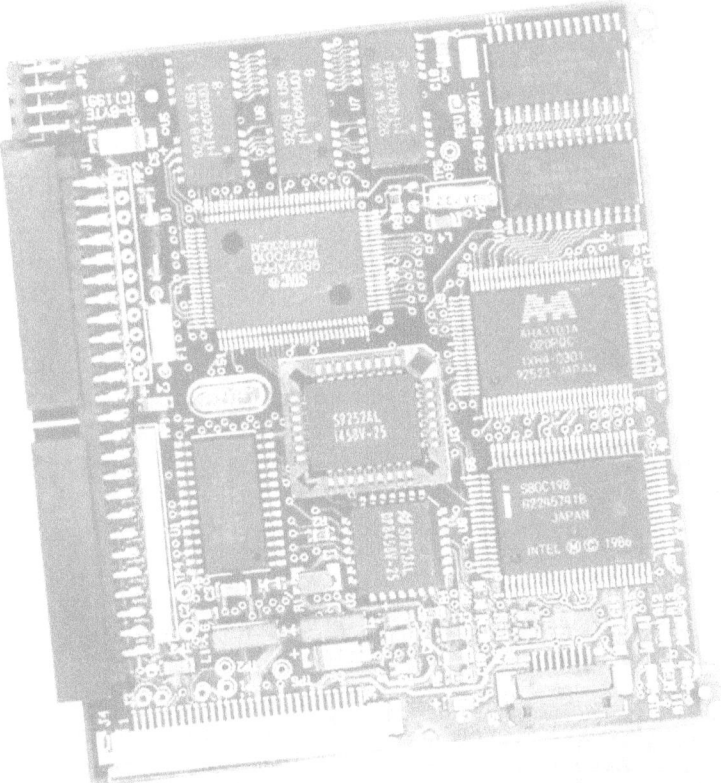

Generating Specifications

A PCA development program begins with writing specifications that define all the requirements the PCA must meet. All departments or functional groups involved with the PCA throughout its lifecycle should participate in generating the specifications. Besides engineering, the following groups should participate:

- Marketing
- Manufacturing
- Materials (also known as purchasing)
- Field Support (may be part of marketing)
- Technical Publications
- Quality Assurance

Though many departments are involved, marketing and engineering drive the specifications because they have the largest stake in the front end of the project.

The project leader or program manager calls a kick-off meeting and each department sends a representative who reports back to the department manager. The department managers usually assign people to the project shortly after the kick-off meeting; this may be when you become involved in the project as the lead design engineer.

After the kick-off meeting, marketing generates the initial specification, called the Marketing Requirements Document (MRD).

Generating the Marketing Requirements Document

TIP
The MRD may also be called the Product Requirements Document (PRD).

Marketing writes an MRD to specify what the product should be and do. The MRD includes the following information:

- Overview of the product
- Target market and market analysis
- Product features, detailing appearance and functionality
- Manufacturing cost estimates and selling price targets
- Schedule targets
- Sales projections
- Return on investment (ROI) analysis

Marketing sends the finished MRD to engineering for review.

Generating the Engineering Requirements Document

TIP

The ERD may also be called the engineering specification or the functional specification.

Engineering responds to marketing's requirements by analyzing the feasibility of the requirements and proposing a technical solution in an Engineering Requirements Document (ERD). As the lead design engineer, you write the ERD. To create an ERD you must perform some preliminary design work. You may also experiment and create breadboards (temporary implementations of some or all of the PCA) to prove that certain concepts work in practice. Electrical, mechanical, software, and systems engineers may need to be involved in this process as well. Manufacturing may also need to be involved, to buy into the feasibility of the project.

Include the following information in the ERD:

- Overview of the product
- External specifications, including:
  - Interface specifications (connectors and cables, interface standards such as USB or Wi-Fi, if applicable)
  - Functions and features
  - Electrical specifications (supply voltage, power consumption, input/output pin characteristics, etc.)
  - Physical specifications (size, weight, markings, etc.)
  - Environmental specifications (operating temperature, storage temperature, shake and vibration)
- Internal specifications, at a high level, including:
  - Block diagram
  - Key components (microprocessors, memory, controller chips, programmable components, etc.)
  - Explanation of operating concepts
  - Software specifications (register addresses and functions, software requirements and hints)
- User guide (any special considerations that effect how the end user uses the product)
- Agency certifications and testing
- Quality and reliability specifications

- Manufacturability and testability considerations
- Cost estimate
- Schedule estimate
- Risks and dependencies

After the ERD is completed, marketing reviews it to ensure it reflects the requirements of the MRD.

Negotiating and Signing off the Specifications

Marketing and engineering negotiate and create new iterations of the MRD and ERD until they agree on a consistent set of specifications. The negotiations are typically conducted via a series of specification review meetings. The meetings should include other involved departments such as manufacturing and technical publications. Involving other departments helps ensure better specifications and begins building a strong project team.

Once a final set of requirements is complete, the functional groups sign off the final specifications indicating their assent. To sign off, a representative of each functional group signs their name on the covers of controlled versions of the final specification documents.

Once the requirements are signed off, it is time to create a project plan.

Planning the Project

A well-led project always has a clear project plan. You need to create the project plan, including the schedule. Show the interdependencies between tasks in the schedule and highlight critical path tasks. Get input from the rest of the project team. The project team needs to review, negotiate and commit to all details of the project plan.

After the project team approves the project plan, you can start developing prototypes.

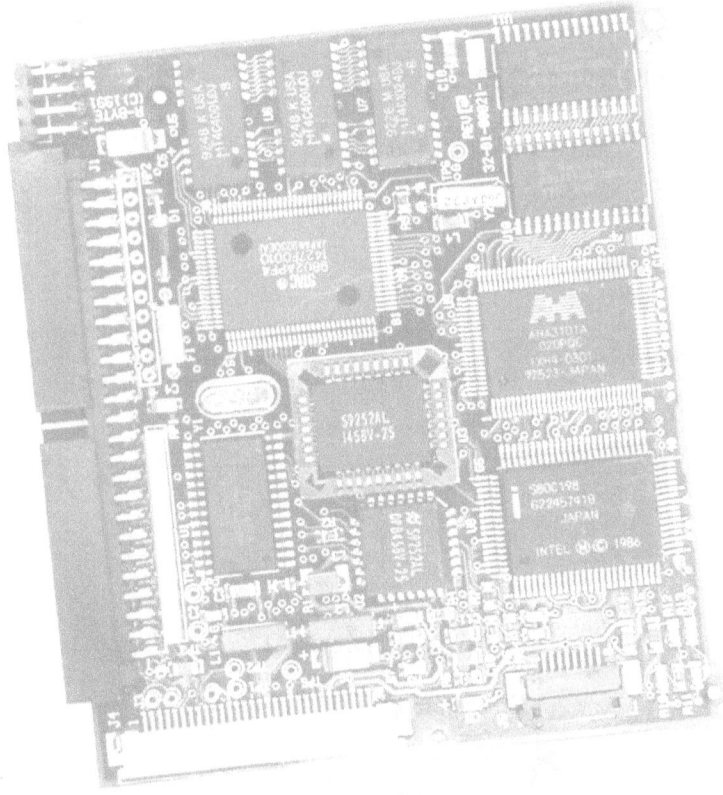

Developing Prototypes

Y ou typically build three revisions of the prototype PCA during the course of a project, successively refining the design. The process for building each prototype is essentially the same. In this section, the overall process of building a prototype PCA is described and then the differences between the three revisions are presented.

Building a Prototype

Each prototype revision involves the following activities:

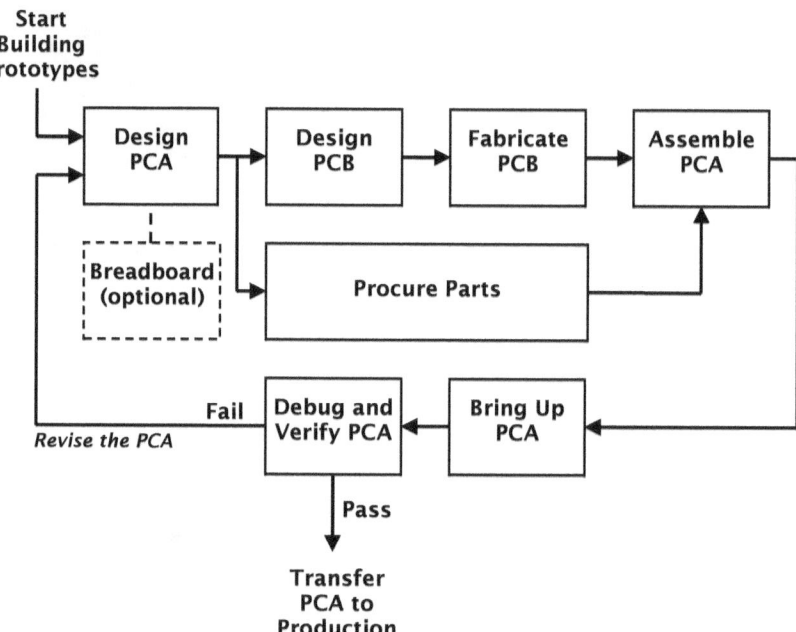

To build three prototype revisions means to go around the loop in the flowchart three times. Note that the parts procurement activity during the second and subsequent revisions is minor because enough parts should be procured in the first revision to cover all the program's anticipated needs. Additional parts procurement activity is only needed when parts are changed or part shortages occur.

Designing the Printed Circuit Assembly

Design a PCA that fulfills the specified requirements by performing the following activities:

- **Select Components.** The components must implement the PCA's specified functions while meeting the size and cost constraints of the project.
- **Draw the Schematic.** The schematic shows how the components are electrically connected on the PCA.
- **Generate the Bill of Materials (BOM).** The BOM specifies the PCA's components by part numbers, descriptions, quantities and locations on the PCB. (See *Glossary* for a complete list of BOM items.)

Note: You must maintain the BOM throughout the project, updating it to reflect all changes.

- **Develop Programming for Programmable Components.** Programmable components are ROMs, PALs, FPGAs, and microcontrollers which require customization via programming.

TIP
Functional tests are needed in production and are sometimes used in prototyping phases.

- **Develop Functional Tests.** Develop a functional test strategy, test procedures, and any fixtures required to functionally test the PCA.
- **Design Mechanical Aspects of the PCA.** Specify the following for the PCA:
    - Exact physical dimensions of the PCB
    - Envelope (height restrictions)
    - Mounting holes and hardware
    - Special hardware like switches and pushbuttons
    - Indicator lights
    - Heat and cooling considerations

If one or more mechanical engineers are assigned to the project, work with them to design the mechanical aspects of the PCA.

TIP
The Theory of Ops Document is sometimes called the internal specification because it explains how the PCA's internal circuitry works.

- Write the Theory of Operations Document (TOPs). The TOPs document is a design review in written form. It contains all the information needed to understand how the board works and how it is built. Users of the TOPs document include engineers in many roles, including manufacturing, test, sustaining, safety, certifications, and software. The TOPs may be written as an update or an elaboration of the ERD.

- Hold One or More Design Reviews. Designing the PCB should not begin until the entire project team signs off the design at a final design review.

Breadboarding Critical Circuits (Optional)

A breadboard is a working model some or all of the PCA, built without the product PCB. Breadboard fabrication methods may include a mix of wire-wrap, point-to-point hand wiring, cobbling together vendor demo boards, or modifying existing PCAs. You can also build a breadboard using an out-of-form-factor PCB that does not conform to the mechanical requirements of the product PCA.

TIP
The "product PCA" refers to the PCA that meets all the design requirements; it is the end product you are designing.

Breadboarding is appropriate if you are uncertain whether particular aspects of the design will work. However, many PCA projects consist of wiring together a given set of very large scale integrated circuits (VLSI chips) per the schematics in their data sheets, along with a power supply and clock circuitry taken directly from manufacturers' application circuits. In such cases, fully functional, in form-factor PCAs are easily achieved on the first prototype round without any breadboarding.

Disadvantages of breadboarding include:

- Effort and funds are expended for something that is not the actual product.

- Breadboarding delay the release of the product.

- Breadboarding may be used to procrastinate, delaying when you confront and solve important challenges of the product PCA such as mechanical constraints, reliability, and usability.

- The product PCA may be left with insufficient capabilities for debugging and software development because the breadboard fulfills these needs at the front end of the project. Then at the back end of the project, when work must switch from the breadboard to the product PCA, lack of these capabilities can cause difficulties.

Designing the Printed Circuit Board

Designing the PCB means to draw the PCB's mechanical outline and the copper traces that interconnect the components of the PCA. Usually a PCB design specialist performs this labor-intensive task, although some engineers prefer to do it themselves. The PCB designer may be an employee of your company or be an outside resource; in either case, the design procedure is the same.

Working with a PCB designer, drive the PCB design process by following these steps:

TIP
PCB designers bid on jobs at a set cost, not an hourly rate; early design work is factored into the overall price.

1. **Involve the PCB designer in early design decisions.** Address the following issues:
   - What design rules to follow, for example, the smallest size allowed for via holes and traces
   - How many routing layers the board will need
   - How close chips can be packed together
   - How to handle certain types of chip packages

TIP
The schematic CAD tool generates the schematic, the partslist, and the netlist.

2. **Give the PCA documentation to the PCB designer.** The PCB designer needs these documents:
   - Schematic
   - Netlist and partslist
   - BOM
   - Dimensioned PCB outline drawing
   - Copies of all the component specifications including package drawings

Give these documents to the PCB designer in hardcopy (in a big three-ring binder) and in machine-readable form (on a CD-R).

3. **Review the documentation with the PCB designer.**

    Alert the designer to any special considerations or concerns you have about the design.

4. **Give the PCB designer a purchase order, if needed.**

    If the PCB designer is an outside resource, a purchase order needs to be issued to pay for the service.

5. **Support the PCB designer as she designs the PCB mechanical outline.**

    Make sure the PCB mechanical outline is complete and correct before proceeding to the next step. The outline includes the external outline and any internal cutouts and mounting holes.

6. **Support the PCB designer as she places the components on the PCB.**

    Do the following to support the designer:

    -   Verify that any special constraints are met. For example, some components' heights or heat dissipation may dictate they go in certain locations.
    -   Help the designer pick the best placement for key components based on functional clustering and interconnect.
    -   Make sure all the connectors have the correct footprints, locations, and orientations.

7. **Support the PCB designer as she routes the PCB.**

    Do the following to support the designer:

    -   Help the PCB designer solve problems.
    -   Make sure critical traces such as clocks and high-speed busses are routed in a way that minimizes noise and crosstalk.
    -   Make sure power and ground distribution is adequate to meet power and noise requirements.

8. **Support the PCB designer as she designs the silkscreen.**

   The silkscreen is the white ink on both sides of the PCB. Make sure the silkscreen shows:

   - Part locations
   - Pin one locations on chips and connectors
   - Polarization of diodes and polarized capacitors
   - Reference designators (the reference designators identify each part, e.g., R2, U12, etc.; the assembly house loads the parts according to these reference designators, which you list in the BOM)
   - PCB part number and revision (may be in copper rather than silkscreen)
   - PCA part number
   - A silkscreen box next to the PCA part number for the assembly house to mark the PCA revision
   - Any other labels or graphics you want on the PCB

9. **Review the final Gerber data (the data files from which the PCB will actually be built) with the PCB designer.**

   This is your last chance to catch errors before the PCB is fabricated, so double-check everything carefully with the PCB designer. Be sure to check:

   - Package footprints and pin one locations on chips and connectors
   - Routing of critical nets like clocks and buses
   - Integrity of power and ground planes, especially if signal traces are cut into the planes and if jumpers are used to connect islands of power and ground
   - Adequate power and ground distribution—check for adequate trace widths on current carrying traces; check for necking down of power and ground planes
   - Thermal reliefs at power and ground vias
   - Solder mask

TIP

Lay the actual parts on printouts of the surface layers to check for proper fit.

Note: The PCB layout program verifies that the interconnection of the parts on the PCB match the netlist, so a manual line check is not necessary.

---

10. **Review the fab drawing and assembly drawing with the PCB designer.**

    At the end of the PCB design process, the designer produces a fab drawing and an assembly drawing. The fab drawing shows the PCB mechanical outline, locations and sizes of all the holes, the layer stack-up, and fabrication instructions. The assembly drawing shows where all the components are placed on the PCB and lists special assembly instructions. Review this information with the designer to verify it is complete and correct.

11. **Tell the PCB designer to transmit the Gerber data and the fab drawing to PCB fabrication house.**

12. **Approve the PCB designer's invoice, if needed.**

    The engineer's approval tells accounts payable the job has been completed and they can pay the vendor.

Fabricating the Printed Circuit Board

Drive the PCB fabrication process by following these steps:

1. **Review the direction of the design with the fabrication house early in the design phase.**

    Address the following issues:

    - The PCB's requirements such as size, number of routing layers, thickness, materials, production quantity, cost, and schedule

    - Design rules such as minimum trace width and spacing, number of traces between pads of dense chips, minimum via size

    - Any special or advanced fabrication techniques such as buried or hidden vias

Facilitate direct communication between the PCB designer and the fabrication house's engineer so they can work out detailed issues together.

2. **Instruct the PCB designer to transmit the final Gerber data and fab drawing to the fabrication house.**

3. **Review the fab drawing with the fabrication house to make sure they understand and agree to every detail.**

4. **Give the fabrication house a purchase order for the job.**

5. **Support the fabrication house while they build the PCBs.**

   The fabrication house usually does not need support once they start the job, but sometimes questions come up. You need resolve any issues quickly to avoid delays in the schedule. Be sure the fabrication house knows you are available and anxious to help so they do not make assumptions.

6. **Receive and inspect the fabricated PCBs.**

   Inspect the PCBs by following the procedure *Inspecting Printed Circuit Boards* in the next section.

7. **Approve the fabrication house's invoice.**

   Your approval as lead engineer tells accounts payable the job has been completed and they can pay the vendor.

The fabrication house visually inspects and SOT (shorts and opens test) tests the PCBs, so the quality of the boards you receive should be high.

Inspecting Printed Circuit Boards

When you receive a set of PCBs from the fabrication house, inspect them using the following procedure:

1. **Remove a PCB from the package.**

   The PCBs arrive bundled together, usually ten to a bundle, wrapped in plastic.

   Note: PCBs are not sensitive to electrostatic discharge (ESD), so you can handle them without wearing a ground strap.

2. **Verify part numbers.**

   Check the following markings:

   - PCB part number and revision level
   - PCA part number
   - Silkscreen rectangle next to the PCA part number for the assembly house to mark the revision

3. **Examine general appearance and workmanship.**

   Check the following:

   - Cleanliness
   - Size
   - Fit of parts
   - Internal layers—present and in correct order

4. **Inspect traces for shorts and opens.**

   Examine the workmanship of the copper traces on the PCB surfaces and check the following:

   - Correct trace widths
   - No shorts (connections between traces) or opens (unintended discontinuities)

   If possible, inspect traces under a microscope. Scan and spot-check the PCB. You do not need to discover every possible defect; just verify that the workmanship is good.

5. **Inspect the solder mask.**

   Verify the following:

   - The solder mask is present.

> **TIP**
> Check fit of parts by placing some parts on the PCB and verifying they fit correctly—especially high pin-count chips and connectors.

- Openings for the component pins align with the pins.
- Solder mask width is adequate to prevent solder bridging between pins.

6. **Check for power and ground shorts and continuity.**
   Using a DVM, verify that the power and ground planes connect to all the required areas and the planes are not shorted together.

7. **Repeat Steps 1 to 6 to inspect several more PCBs.**
   If you find any problems, work with the fabrication house to resolve them. You must make a decision about whether work can continue with these PCBs, or whether they should be discarded and new PCBs should be fabricated.

If you do not find any problems when you inspect the PCBs, you can assume the whole lot is good. Send the PCBs to the assembly house to assemble the PCAs.

Procuring Parts

Obtaining the components that go on the PCB often dictates the critical path of the schedule because many parts are difficult to obtain or have long lead times. Usually the purchasing department assigns a buyer to support engineering to procure parts. In addition, a senior buyer or materials manager participates on the project team to deal with materials issues. In other cases, an engineering technician procures parts. Some large companies have engineering services departments that procure parts, get PCBs fabricated, and assemble prototypes, greatly relieving your work load. At the other extreme, sometimes you need to procure the parts yourself. No matter who does it, parts procurement is a time-consuming and exacting job.

To drive the parts procurement process, follow these steps:

1. **Generate the bill of materials.**
   The BOM should indicate the number of assemblies for which parts should be purchased, factoring in extra

parts needed for loss, damage, experimentation, and PCA repair. Enough parts are purchased at one time to supply the needs of all the prototyping revisions because purchasing is so much work and because many types of parts are difficult to obtain in small quantities.

2. **Discuss parts strategy with the PCA assembly house.**

   The PCA assembly house may be able to supply certain standard components such as resistors and capacitors, eliminating the need to purchase these parts. This service can be a great help because you may only need a small quantity of a certain part, but that part may only be available in large quantities. For example, resistors and capacitors usually come packaged on reels of three thousand or five thousand pieces, which is likely to be much more than you need.

3. **Review the BOM with the buyer.**

   Point out any concerns or special problems and make sure the buyer understands all the information on the BOM. The BOM must clearly indicate which parts the buyer should and should not purchase. Parts which are listed on the BOM but which the buyer should not purchase include:

   • No-loads, which are parts that are not needed on the PCA, but which have place-holders on the PCB

   • Parts which will be supplied by the assembly house

   • Parts which you will supply; for example, samples of new parts you obtain directly from the manufacturers of the parts

4. **Work with the buyer to identify long lead-time items.**

   Prioritize items with long lead times to make sure they do not jeopardize the schedule. Work with the buyer to solve lead-time problems by finding alternate sources or alternate parts. You may change the design to avoid parts with availability problems, or you may elect to use alternate parts for early prototypes.

5. **Support the buyer as she orders the parts.**

   The buyer orders all the parts, giving priority to parts that have long lead times.

6. **Supervise kitting.**

   Parts are collected in a box as they are received. The box of parts is called the kit for the PCA. The buyer tracks the arrival of all parts so she always has a record of what is in hand and what is on order. The buyer or the engineering technician takes physical possession of and responsibility for the kit.

7. **Audit the kit.**

   As the time for PCA assembly approaches, physically inspect the kit to verify its contents. You do not need to count the individual parts because they arrive from vendors in bags or on reels marked with a quantity. The auditor, usually the engineering technician, simply identifies each part, checks it quantity, and records it on a copy of the BOM. Parts shortages are identified and plans are made to deal with them, such as expediting delivery, finding alternate sources, or deciding to assemble the PCAs without certain parts. Auditing the kit also establishes a firm baseline to refer to when the PCA assembly house comes up short on some parts.

8. **Transfer the kit to the PCA assembly house.**

   Transfer the kit two weeks before the start of PCA assembly to provide time for the PCA assembly house to audit the kit.

9. **Support the PCA assembly house as they audit the kit.**

   Work with the PCA assembly house to resolve any parts shortages they encounter when they audit the kit and assemble the PCAs.

Assembling the Printed Circuit Assembly

Drive the PCA assembly process by following these steps:

1. Involve the assembly house in early design decisions.

2. Transmit the BOM to the assembly house and work with the assembly house to resolve BOM issues.

3. Transmit the assembly drawing to the assembly house and review it with the assembly house's engineer.

4. Give the assembly house programming files and instructions for any programmable components you want them to program.

5. Deliver the parts kit to the assembly house.

6. Deliver the PCBs to the assembly house.

7. Give the assembly house a signed purchase order for the assembly job.

8. Give the assembly house the go-ahead to build the PCAs.

    Be certain the assembly house has the correct, final BOM to work from.

9. Support the assembly house while they build the PCAs.

    The assembly house may encounter problems and have questions as they build the PCA. To avoid delays in the schedule, you need to be available to quickly resolve any issues that arise. Be sure the assembly house knows you are available and anxious to help so they do not make assumptions.

10. Receive and inspect the completed PCAs.

    Inspect the PCAs by following the procedure *Inspecting Printed Circuit Assemblies* in the next section.

11. Approve the assembly house's invoice.

    Your approval as lead engineer tells accounts payable the job has been completed and they can pay the vendor.

The assembly house SOT tests the PCAs and also tests that the correct passive components (resistors and capacitors) and most active components (diodes, transistors, and FETs) have been loaded. The assembly house generates these tests from the netlist you provided to them. If you want the assembly house to run one or more functional tests on the PCAs, then do the following:

1. Develop functional test procedures.

2. Provide any necessary fixtures to the assembly house.

3. Train the assembly house's personnel on how to run the tests.

Note: A functional test is usually put in place for production but not for prototyping.

Inspecting Printed Circuit Assemblies

When a set of PCAs is received from the assembly house, inspect them using the following procedure:

1. Open the package and remove a PCA.

    The PCAs arrive in anti-static protective wrapping, usually clear pink bubble-wrap or clear pink plastic bags.

    Caution: Everyone handling PCAs must wear ground straps to prevent damaging components through discharge of static electricity (ESD).

2. Examine general appearance and workmanship.

    Is it the correct PCA? Is it clean? Do most of the parts appear to be present—on both sides of the PCA? Did the assembly house mark the PCA revision?

3. Verify component stuffing.

    Verify that all the components that are supposed to be on the PCA are present, and all the components that are supposed to be no-loads are absent.

4. **Verify component orientation.**

    Verify that all the components are installed in their correct orientations. Pay particular attention to VLSI chips, connectors, polarized capacitors, and diodes.

5. **Examine workmanship of solder joints.**

    Use a microscope to verify that the solder joints are well formed, the component pins are secured to the pads, and solder does not bridge between pins.

6. **Verify that programmable components are correctly programmed.**

    If the PCA has programmable components that are programmed and installed by the assembly house, check that they are marked appropriately to indicate they have been programmed with the correct revision of code.

7. **Check for power and ground shorts and continuity.**

    Use a DVM to verify that the power and ground planes connect to all the areas they need to go to, and that there are no shorts between the planes.

8. **Repeat Steps 1 to 7 to inspect several more PCAs.**

    If you do not find any problems after inspecting several PCAs, you can assume the whole lot is good.

If you do not find any problems when you inspect the PCAs, you are ready to start bring-up. If you do find problems, work with the assembly house to resolve them. No matter how bad the problems are, you should still be able to repair the PCAs, even if a heroic amount of rework is needed. The assembly house may perform rework for free if the problems are due to negligence or errors on their part. An engineering technician or engineer can perform the rework in-house, if the rework is not extensive.

Bringing Up the Printed Circuit Assembly

Bringing up the PCA means to power it up and get it working. PCA bring-up is complete when the PCA is functional enough to be used for another purpose such as software development or product demonstration. A PCA which has passed bring-up but not debug and verification (see next procedure) is assumed to have many problems yet to be discovered and fixed; use of these PCAs is on an "as is" basis. Nevertheless, PCA bring-up is a major milestone because even a partially working board strongly indicates the design, fabrication, and assembly of the PCA were successful and the prototypes are able to serve their purpose.

Bring the PCA up to a basic level of functionality using the following steps:

1. **Prepare a power supply to power the board.**

   Set up the power supply to generate the proper voltages needed by the PCA. The power supply should be laboratory quality and current-limiting so the PCA is not damaged if it has defects that allow excessive current to flow. The types of defects that could cause this condition include power to ground shorts and diodes or polarized capacitors installed backwards.

2. **Connect a PCA to the power supply.**

3. **Turn on the power.**

   If smoke appears or if the power supply current limits, it indicates excessive current flow caused by the type of defects discussed in Step 1. Repair these defects before proceeding to the next step.

TIP
Engineers jokingly call powering up the PCA for the first time "letting the smoke out of the board" because smoke often appears.

Caution: If smoke appears or if the power supply's current limiting circuits are triggered, turn the power off immediately to prevent a fire from igniting and to minimize damage to the PCA.

4. **Feel for hot spots.**

    With power applied, move your hand over the PCA to see if any areas are generating excessive heat. If you do not feel excessive heat, touch some or all of the components with a fingertip to feel if any of them are getting hot. If any components feel hotter than they should normally be, it indicates problems that should be investigated and fixed before proceeding to the next step.

5. **Verify the quality of the power supplies on the PCA.**

    Use a DVM and an oscilloscope to measure the voltages and examine the noise of any power supplies on the PCA. If the voltages are not within the tolerances specified for the design, or if they are excessively noisy, it indicates problems that should be investigated and fixed before proceeding to the next step.

6. **Verify the quality of the clock signals.**

    Use an oscilloscope to examine the frequencies and waveform shapes of the PCA's clock signals. If the frequencies are incorrect or the waveform shapes are not of high quality, it indicates problems that should be investigated and fixed before proceeding to the next step.

7. **Test the functionality of the PCA's core circuitry.**

    Most PCAs have a set of core circuitry that must work before the rest of the PCA can work. Often the core circuitry is a microprocessor and associated memory chips that must work for the PCA to run its software. Bring up this core circuitry first, starting with the most basic functions and working toward the more complex functions. Additional equipment may be needed at this time; for example, you may need to hook up a microprocessor emulator, a display, or a desktop computer. You may ask a software engineer to work with you at this point.

8. **Test the functionality of the rest of the PCA's essential circuitry.**

   Once the core circuitry is functioning, bring up the rest of the PCA's essential circuitry. Work on one functional module at a time. Functional circuitry is essential if it is needed for the PCA to serve its next purpose; for example, if a software developer will use the PCA, only the circuitry that developer will need is essential.

If you encounter problems at any step in the bring-up process and find it difficult to proceed to the next step, set the PCA aside and try to bring up a different PCA. During prototyping, a number of PCAs may have unique, serious fabrication or assembly defects that are very difficult to analyze and repair. Setting aside such PCAs and making progress on better ones is a good strategy. For this reason, assembling a single PCA is never wise. If a single PCA is bad, you have no way to determine whether it is simply a rogue PCA, or the failure is a systemic problem in the design, fabrication, or assembly. A minimum of three PCAs should always be available for PCA bring-up.

TIP
Serialize the PCAs and keep a log of each unit's problems and progress; also where they are and who is responsible for them.

After PCA bring-up, distribute prototypes per the product plan to support various activities such as software development and marketing demonstrations. You should retain a number of models to bring up any functionality not yet achieved, and to find all the additional problems in the design and in the fabrication and assembly processes.

Debugging and Verifying

Debugging means finding and fixing problems (bugs). Design verification is testing the prototype PCAs to prove they meet every specification set forth in the MRD and ERD including functionality, appearance, fit, and power consumption. Electrical, mechanical, and software engineers should each verify the PCA meets the requirements in their areas of expertise. Prepare a design verification test plan in advance of verification testing, preferably as part of the project planning process; and publish a design verification report at the end of each prototyping cycle.

RF, EMI, and ESD problems can be very difficult to diagnose and fix. Start preliminary testing in these areas during the first prototype build if possible, so subsequent builds can converge on solutions to these problems. RF, EMI, and ESD test facilities and personnel must be scheduled ahead of time because they are usually on tight allocation. You might ask specialized experts to run these tests and help solve the problems that are uncovered.

Note: Reliability testing to verify the mean time between failures (MTBF) specification must be performed on product produced on the manufacturing line, so this testing cannot begin until pilot production. Reliability testing takes three months to a year, requires a substantial number of units, and is often performed by manufacturing or quality assurance instead of engineering.

During debugging and design verification, you may need to quickly turn (re-design and fabricate) some PCBs to test fixes like rearranging ground planes or adding shielding. Such turns are strictly experimental and should not be counted against the three-revision project goal. These experimental PCAs should never leave the engineering lab; nevertheless, they should be marked and documented as formal revisions for identification and control purposes.

Each prototype build cycle is costly and time consuming. Therefore, do as much debugging and design verification as possible on each round of prototypes before revising the PCB and building another round. Some engineers make the mistake of turning a new PCB revision as soon as they discover a serious problem or a handful of problems. This approach is costly in time and money and leads to a great deal of confusion trying to manage the different revisions of the PCB. Getting as much use as possible out of a single revision of the PCB, even if extensive rework is needed, is better than building a lot of extra PCB revisions.

Developing Three Revisions

Most well run PCA programs can be accomplished with only three revisions of the PCB. The following flowchart provides an overview of the three revisions:

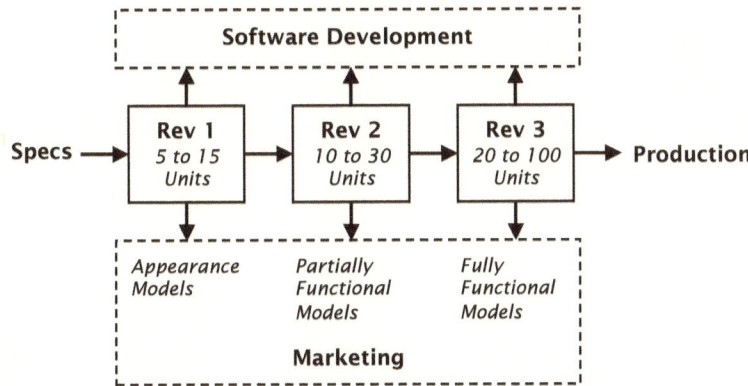

First Revision. Rev 1 PCAs are made to be completely or mostly functional with some amount of rework. Heroic amounts of rework may be applied if necessary, because learning as much as possible from these prototypes is essential. Only a small number (five to fifteen) of Rev 1 PCAs are built because they are likely to have problems that make them unusable for most purposes. Some of these models are used by software engineering to begin developing the product's software. Some of these models are given to marketing as non-functional, appearance-only models to show to key customers.

Second Revision. Rev 2 PCAs are made to be completely functional with only a small amount of rework. Only a modest number (ten to thirty) of Rev 2 PCAs are built because supporting them consumes engineering resources. These models are used for software development and extensive hardware verification. Some of these PCAs are given to marketing as partially functional models to share with key customers; some customers may be permitted to start testing and qualifying the product with these models.

Third Revision. Rev 3 PCAs are made to be completely functional with only minor rework. Only a limited number (twenty to one hundred) of Rev 3 PCAs are built because supporting them consumes engineering resources. These models are used for software development and extensive hardware verification. Some of these PCAs are given to marketing to place with key customers so they can test and qualify the product for their application.

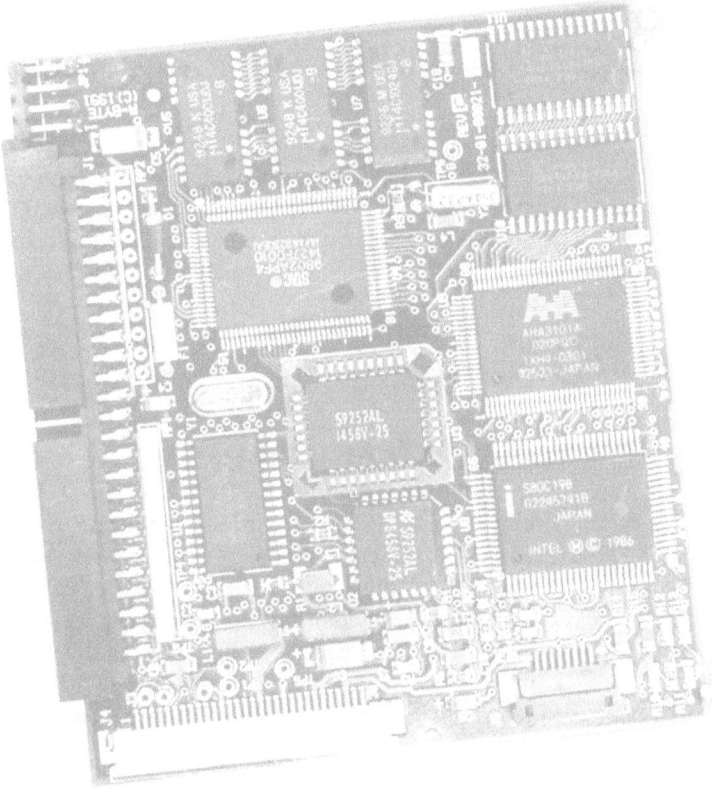

Going to Production

A fter the PCA has passed design verification, you are
ready to certify it and release it to manufacturing
for mass production. The process to take a PCA
from Rev 3 prototypes to production consists of
the following steps:

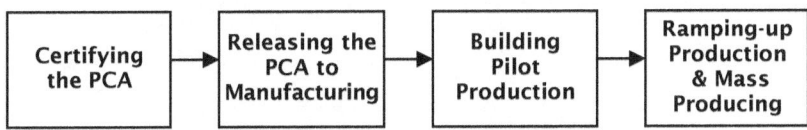

| Certifying the PCA | → | Releasing the PCA to Manufacturing | → | Building Pilot Production | → | Ramping-up Production & Mass Producing |

35

Certifying the Printed Circuit Assembly

Products sold commercially are legally required to be certified for safety and other factors like RF interference by agencies such as UL, CSA, and the FCC. Certification testing is usually done using Rev 3 prototypes, but some certifications must be tested, or at least regression tested, on production product.

Releasing to Manufacturing

The product documentation is released for production through the company's document control process. Responsibility for the product is formally handed off from engineering to manufacturing at a Transfer of Information (TOI) meeting. However, manufacturing personnel work with engineering as a key part of the development team from the start of the project, and engineering continues to support manufacturing after the transfer is complete.

Building Pilot Production

The manufacturing department builds their first round of PCAs. As the design engineer, you continue assisting in bringing up PCAs and debugging the production process. You train the manufacturing engineer, and possibly the line workers as well, to run the functional tests you developed, and you debug any problems with these tests.

Ramping-up Production and Mass Producing

The manufacturing department gradually increases the rate at which they build PCAs until full mass production is reached. Your involvement as the design engineer becomes less and less as manufacturing takes over full responsibility for the PCA.

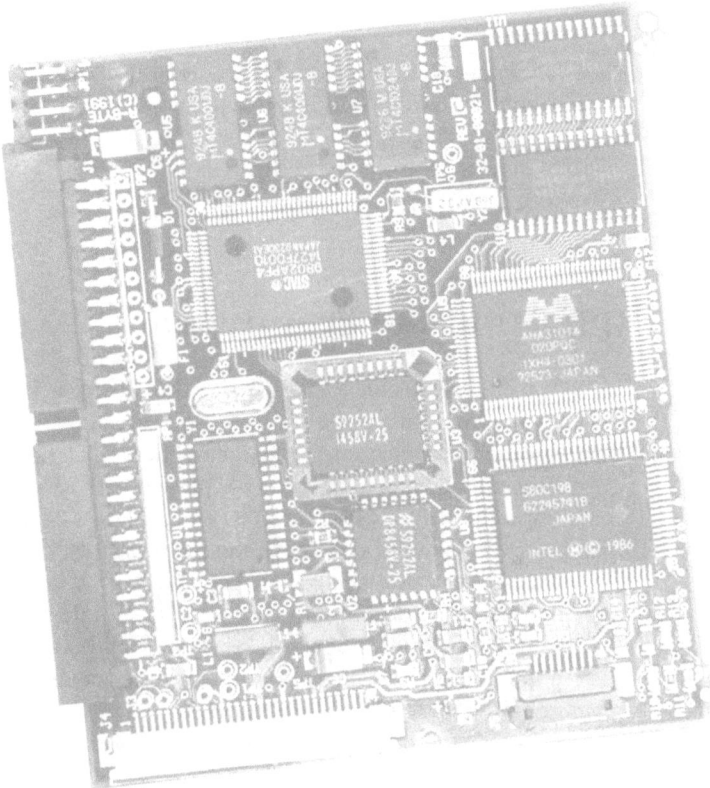

Cross-Functional Support

T hroughout the project you interact with and support other departments in their work to develop, market, and manufacture the PCA. This section discusses the activities this support entails.

Participating in Program Meetings

Throughout the project, the project leader or the program manager conducts weekly program review or status meetings to coordinate actions among all the departments and to track action items. If a program manger conducts the meetings, you should still attend and be prepared to give status on your progress.

Supporting the Prototypes

You are responsible for supporting the prototype PCAs as they are used in house (for example in software development) and externally (for example, as marketing demonstration models). When a model breaks, fix it or replace it with a working one. This responsibility becomes a larger and larger commitment of your time and effort as more and more prototypes are built and released for various purposes. You should minimize the number of models you build at each stage of the program so your bandwidth is not totally consumed by this support function. The program plan must include enough engineering resources to provide this support. A good technician is essential, and often a junior engineer is assigned to assist in this responsibility.

Supporting Software Development

Most PCAs require software to implement their functionality. The PCA hardware engineer must work with the software engineers to help them understand the PCA and write software for it. Hardware engineers and software engineers need to work closely together to solve problems, regardless of whether the problems are caused by hardware or software.

Supporting the Technical Writer

Include the technical writer in the project from the beginning, while the specifications are being generated. Work with the writer to develop a documentation plan for the PCA as part of the overall project plan. As the design engineer, you need to keep the technical writer informed of any changes in the design and give the writer the information and training needed to successfully produce high quality product documentation.

Supporting Marketing and the Field

Support for the marketing, sales, and field support groups by performing the following activities:

- Negotiating product specifications
- Providing early mock-ups for marketing to show to customers
- Providing appearance and functional models for marketing to share with selected customers
- Keeping the functional models running
- Going on sales calls to share information with key customers while the product is still being developed
- Solving customers' technical problems when the technical support group cannot handle them because the problems are too difficult or because the product has not yet been transferred to technical support

Supporting the Product at End of Life

Engineering may need to support manufacturing when a product reaches end of life due to technical causes. For example, one or more components used on the PCA may become unavailable for one reason or another, or perhaps their cost or lead-times become exorbitant. In some cases the manufacturing department may be able to find alternate components on their own. In other cases, the PCA's design engineer (or another design engineer assigned to support the PCA) may be brought into the process to qualify alternate components or to suggest solutions. If no solution can be found, the PCA can no longer be manufactured and therefore has reached the end of its life. The product management team should have a plan in place to deal with the end of life of a PCA. A new product may be available to take its place; however, buffer stock may need to be warehoused and customers may need transition plans. You should be alert to possible problems that may cause the PCA to reach end of life, and you should inform management of any risks you foresee.

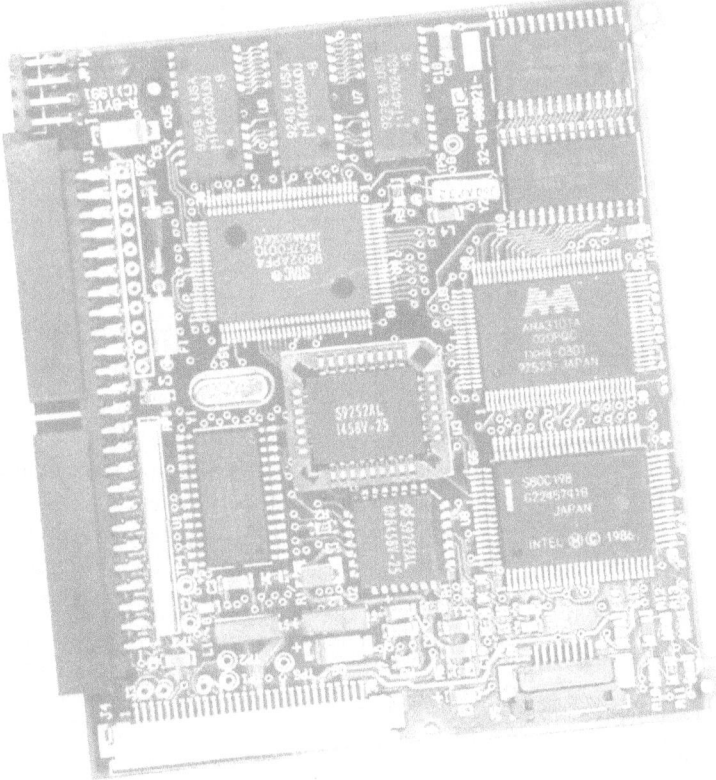

Troubleshooting

T his section provides advice for dealing with common problems that occur during a PCA development program.

Marketing Won't Write a Marketing Requirements Document

Marketing may resist writing an MRD for the following reasons:

- Marketing personnel are too busy and the project is a low priority for them.
- The project originates from within engineering and is not officially approved outside of engineering.
- The company has a poor development process.

To revolve this problem, write the MRD yourself and ask marketing for comments. Marketing personnel are usually happy to give feedback on something you show them, even if they do not have time to work on it themselves. You are likely to get better feedback if you present them a marketing-oriented document or a physical model rather than an engineering specification. A good approach to get started is to mock up a product brief.

Specifications Won't Settle

Sometimes negotiations of marketing requirements and engineering specs go on forever. You can give the players an incentive to reach agreement by insisting on slipping the program schedule, explaining that you cannot complete the design until the specs are finalized. Do not try to stay on schedule without firm specs because you will wind up modifying the design later on, and the schedule will slip anyway—probably even more than if you had waited for firm specs.

Software Engineers Won't Review the Hardware Specs

Software engineers do not get serious about a piece of hardware until a working sample is put in their hands. This behavior is not due to a lack of willingness, but rather because software engineers are busy people, because they do not appreciate the cost of changing hardware later on, and because they assume you understand their needs. You should make every effort to get software engineers to treat the specs seriously at the front end of the project. Nevertheless, to be successful, you need to become an expert in software yourself and take responsibility for designing hardware that is a pleasure for software engineers to work with. Do not hesitate to visit software engineers' cubicles to ask for ideas about design decisions that will eventually affect their lives, or to ask for help with software-related issues you are not sure about.

Hardware and Software Engineers Pointing Fingers at Each Other

"It's a software problem; I'm not working on it." "No, it's a hardware problem and I can't do my job until the hardware is fixed!" This behavior never fixes problems. Whenever the source of a problem is in doubt, you, as the hardware engineer, should assume the problem is in the hardware and work on fixing it from that perspective. Simultaneously, software engineers should assume the problem is in the software and work to fix the problem from that perspective. Moreover, hardware and software engineers should work together, shoulder to shoulder on the bench, to solve difficult problems.

Can't Get Parts

Do your best to design with parts that are available and have feasible lead times. In large manufacturing companies, picking parts that are already qualified and used in production of other products is usually a safe approach. Some engineers in smaller companies choose parts directly from the Digi-Key catalog because they know major distributor Digi-Key can deliver small quantities of parts in a timely fashion to support prototype builds. When the buyer runs into parts availability problems while placing orders, the engineer is often called upon to help find solutions. Sometimes you can find a distributor who can supply the parts or you can obtain samples directly from a part's manufacturer. In other cases you may need to find an alternate part or change the design to use a different part. For example, if a 33.2k resistor in a 0402 package is unavailable, a 33.2k resistor in a 0603 package can easily be substituted. You might change the design to use the 0603; or, if the 0402 will become available later on, you might leave the 0402 in the design and instruct the assembly house to use the larger part for the prototypes.

Prototypes Have Problems

Sometimes a set of prototypes has problems in the design or in the manufacturing process that makes the whole lot unusable. Although this kind of disaster should be a rare occurrence if you do your job right, here are some suggestions for handling this unfortunate situation:

- **Do Not Panic.** PCA problems are not the end of the world. Stay professional. Act patiently and intelligently.

- **Do Not Try To Hide the Situation.** Be forthright with your manager and project team. Explain what you know, what you do not know, and what you plan to do next. Ask for help and support, and offer your support to help others cope with the ramifications of the disaster.

- **Give Schedule Estimates.** The disaster is going to cause the program schedule to slip. Everybody knows it and wants to know how big the slip will be. "I don't know," is probably an honest answer, but it is not satisfactory and does not help them plan. You should give a best estimate and tell everybody you will keep them informed as you learn more. Keep them informed with regular status updates via e-mail, even if the status is that you have not made any progress.

- **Do Not Rush To Turn a New PCB.** Do not revise the design and build new PCBs until you are certain you understand the root cause of the problem and are confident of the solution. You should get all the information you can out of the broken PCA before you start the next revision.

- **Plan an Additional PCB Turn**. Change the project plan to schedule a new PCB revision that fixes the problem. Trying to force the fix into the existing schedule in an ad hoc manner does not work and inevitably leads to worse schedule slips later on.

ESD and EMI Problems

ESD and EMI are notoriously difficult areas to conquer. Be conservative in the design up front, using techniques such as:

- Adding additional protection and noise suppression devices
- Being very careful about ground planes and grounding
- Planning for EMI shielding and gaskets

Be generous about adding extra parts at the beginning of the design because removing unneeded (or costly) components from the design later on is much easier than adding new components to solve problems at the last minute. Try to get a look at ESD and EMI compliance in the early prototype stages to gauge whether the design has problems in these areas. If available, specialists in these areas can assist you with design and testing for these considerations. Talk about these issues at design reviews. If ESD and EMI failures are still present at the final prototyping stage and they are threatening to delay the program schedule, the team should not hesitate to throw some ugly shielding or protective coatings at the problem. People will scream about the impact on the product's manufacturing cost, but the impact of delaying product introduction is usually a lot worse.

Manufacturing Wants to Change the Design

Manufacturing departments are good at what they do. If they want to change something, they usually have a good reason for it. Once the product has transferred to manufacturing, it is not your responsibility anymore. Manufacturing's job is to produce the product while meeting cost, schedule, and quality requirements. Let them make the changes they want. You should cheerfully assist manufacturing when they request help, while complying with the company's procedures and implementing changes via the ECO process.

Management Imposes an Unrealistic Schedule

If the project schedule is unrealistic for any reason, do your best to meet it, but do not cut corners. Stay professional and disciplined; cutting corners catches up on you eventually. Give management realistic feedback about the project's progress, but do not waste energy telling them, "I warned you!" If they do not already realize that, telling them will not help. If you are pressured to work long days and weekends to meet an unrealistic schedule, seriously consider how much of your personal life you are willing to sacrifice for the benefit of the company.

Taking Over a Poorly Run Project

You may be asked to take over an ongoing project. If you are not satisfied with the way the program has been planned, or its current state, do not lower your standards to those of the program. Make it your business to work towards bringing things up to professional standards. You may not be able to make changes abruptly. Cooperate with the rest of the team. Do not confront them but be an advocate for best program management practices. You are working for the greatest success of the project, the team, and the company—not to mention your own career!

Glossary

This section defines phrases and abbreviations used by PCA developers.

Agency Approvals and Certifications

Approvals from UL, CSA, FCC, etc.; products sold commercially must meet certain requirements specified by government and third party bodies; agency approvals and certifications show the product meets these requirements.

Appearance Model

Mock-up of a product that looks like the real thing (to a greater or lesser extent) but does not function; for example, the original Palm Pilot was famously financed based on an appearance model that was simply a wooden block that showed its size.

Assembly

The process of attaching (mostly by soldering) all the required components to the PCB.

BOM

Bill of materials; a table identifying all the parts used in the PCA along with this supporting information:

- Brief descriptions of the parts
- Internal company part numbers
- Manufacturers part numbers
- Quantities and costs or cost estimates

- Locations on the PCA (reference designators)
- Names of manufacturers and distributors
- Lead times
- Order and inventory status

Breadboard

A temporary implementation of some or all of a design; used to verify design concepts before finalizing the design of the product PCA.

Buried Via

A via that connects internal layers of a PCB without connecting to the surface layers.

Capacitors

A type of discrete electrical component that has two pins or contacts.

Chip

A type of electrical component used on PCAs; a semiconductor integrated circuit (IC); usually looks like a little (an inch or smaller) black plastic or ceramic box with protruding metal feet or solder bumps on the bottom.

Clock Circuitry

Circuitry consisting of crystals, transistors, resistors, and capacitors that generates clock signals; clocks are regular, repeating signals that set the timing of chips, for example, the 2.2 GHz clock for a Pentium processor.

Components

Anything that gets mounted on PCBs, including discrete, passive electrical components like resistors and capacitors, active components like transistors and chips, and mechanical components like switches and heat sinks; also called parts.

Data Sheet

A document containing all of the specifications and functions of a component; data sheets give engineers all the data they need when designing the component into a PCA; the data sheet may be a single page for a transistor or a sixty page manual for a complex communications IC.

Debug

The process of identifying, understanding, and fixing problems (bugs).

Diode

A type of discrete electrical component that has two pins or contacts.

Discrete

An electrical component that consists of a single device like a resistor, diode, or transistor; as opposed to an integrated circuit which integrates many devices in a single component.

Document Control

The process which identifies, distributes, and archives official versions of documents and their revisions so all parties know which documents contain current, reliable information; a *controlled document* is a document controlled by the Document Control Department.

DVM

Digital volt meter; a piece of lab equipment for measuring voltage, current, and resistance.

ECO

Engineering change order; a controlled document that specifies changes to be applied to a PCA's design.

EMI

Electromagnetic interference; similar to RF except EMI does not do anything useful.

ESD

Electrostatic discharge; ESD causes the shock you may receive after petting a cat; ESD can disrupt or damage electronic equipment if the circuitry is not designed with adequate ESD protection.

Fab Drawing

Short for *fabrication drawing*; a mechanical drawing specifying the PCB mechanical outline, locations and sizes of all the holes, the layer stack-up, and fabrication instructions.

Fabrication House

A vendor who manufactures PCBs.

FET

Field effect transistor; a type of discrete electrical component that has three pins or contacts.

Firmware Engineer

A programmer who writes low level code for a microcontroller; a firmware engineer knows a little of both hardware engineering and software engineering, which accounts for the derivation of the term *firmware*.

First Article

The initial sample of a fabricated part such as a PCB; the customer may wish to inspect and approve the first article before the manufacturer fabricates more parts.

FPGA

Field Programmable Gate Array; a large customizable logic chip.

Functional Test

A test which operates a PCA in its normal mode, usually connected to a fixture simulating the PCA's application environment; as opposed to SOT and component testing.

Gerber Data

Gerber data or Gerber files refer to a specific file format used for PCB design artwork; the Gerber data is transferred from the PCB designer to the fabrication house to be used in fabrication of the PCB.

Ground

The electrical potential used as the reference for all other voltages; literally the potential of the earth; the minus side of a battery or the third prong of an AC power plug is usually ground.

Ground Strap

A wire that connects a wrist band to an electrical ground to prevent ESD buildup in the person wearing the wrist band; a person handling electronic parts such as PCAs and components wears a ground strap to avoid damaging the parts with ESD.

Hardware Engineer

An electrical engineer responsible for designing PCAs; *hardware* refers to physical objects like electronic components, PCBs, etc.; also called a design engineer or a PCB engineer.

Hidden Via

A via positioned underneath a component; a hidden via is not exposed for inspection or probing from one or both sides of the PCA.

Kit

A package containing a set of parts for a given number of PCAs; kitting means to put a kit together.

Layer Stack-up

Describes the order of the copper trace layers on a multi-layer PCB.

Microcontroller

A small computer on a chip, often customizable via programming.

Model

See *Prototype*.

Multi-layer Board

A PCB with more than two layers of copper traces; two layers are on the two surfaces of the PCB; the additional layers are internal to the PCB; a multi-layer PCB is fabricated by laminating together two or more two-layer boards.

Netlist and Partslist

Data files which are text listings that capture all the information contained in a schematic; the netlist specifies the interconnection of the components, while the partslist identifies the components, their type, and their PCB footprints; schematic entry CAD programs have commands to produce the netlist and partslist files in various formats for transfer to different PCB design programs.

No-load

A part that is not supposed to be assembled (loaded) onto the PCA; no-loads are useful to provide options in the design or to hold back parts to be loaded later because of availability or other reasons.

Oscilloscope

A piece of lab equipment that displays electrical waveforms on a screen; also called a *scope*.

Out-of-form-factor

Does not conform to the required physical dimensions.

PAL

Programmable array logic; a small customizable logic chip.

PCA

Printed circuit assembly, also called a PC board or simply a board; a PCB loaded with components.

PCB

Printed circuit board, also called a bare board (bare of components) or a fab (from *fabricated part*); typically a flat, rigid, fiberglass board with electronic circuit traces etched in copper on the top, bottom, and internal surfaces.

Programmable Components

Components that are customizable via programming, such as ROMs, PALs, FPGAs, and some microcontrollers.

Product Brief

A one or two page summary of a product's features, used as a sales aid; usually very nice looking full color glossy prints.

Project Binder

A three-ring binder holding all the project's PCA-related documentation including specifications, meeting minutes, schematics, BOM, mechanical drawings, and component specifications.

Prototype

A PCA (or a higher level assembly) built by engineering, not by manufacturing; also call a model.

Purchase Order

A legal contract from one company agreeing to pay for goods or services ordered from another company.

Resistors

A type of discrete electrical component that has two pins or contacts.

Rework

Changes applied by hand to a PCA after it has been assembled; rework commonly includes operations such as replacing a component with a different one, cutting traces, and adding jumper wires.

Rework Instructions

Documentation generated by the engineer to specify exactly what rework should be applied to a PCA.

RF

Radio frequency; radio waves which may be used to transmit information to radios, wireless phones, televisions, and computers.

ROM

Read only memory; a customizable memory chip.

Schematic

A drawing that abstractly shows the interconnection of the electronic components of a PCA.

Silkscreen

White ink on the surfaces of PCBs used to indicate and identify component locations, part numbers, etc.

SMT

Surface mount technology; PCAs with electronic components soldered to the PCBs' surfaces, as opposed to thru-hole technology where the pins of the components are inserted into holes drilled through the PCB.

Solder mask

A clear coating covering parts of the surfaces of PCBs to prevent solder from reaching the copper beneath the solder mask; the solder mask has openings to allow soldering where the component pins connect to the PCB.

Software Engineer

A programmer who writes the code for a computer or microprocessor; the term software derives from the fact code is not hardware and it is easy to change.

SOT

Shorts and opens test; a short or short circuit occurs when two signals accidentally connect; an open or open circuit occurs when a signal does not connect to all the places it should, for example because a wire is broken or a copper trace has an unwanted gap.

Specification

A document that describes in detail the requirements or functions of a product, PCA, or component. *Specs* is often used as shorthand for specifications.

Thermal relief

Cutout around a power or ground via to limit the thermal conductivity to the planes when soldering; if the via has a solid connection to the plane (that is, it has no thermal relief), heating up the via sufficiently to allow solder to flow is difficult.

Trace

A copper wire etched on the surface of a PCB, used to interconnect electrical components on a PCA.

Transistor

A type of discrete electrical component that has three pins or contacts.

Turn

Turning a PCB, or a board turn, means to revise the PCB and build samples of the new version.

USB

Universal Serial Bus; an interface between computers and devices like keyboards, printers and digital cameras.

Via

A hole in a PCB, plated with metal, used to connect traces on different layers; a standard via is drilled through all the layers of the board and is accessible for inspection or probing at both surface layers; see also *Buried Via* and *Hidden Via*.

VLSI Chip

Very large scale integrated circuit

Wi-Fi

Short for "wireless fidelity"; refers to certain types of wireless local area networks based on specifications in the 802.11 family of standards.

Index

About the Authors

Elaine Rhodes

developed printed circuit
assemblies and integrated
circuits for twenty-five years
in Silicon Valley companies
like Intel, Tandem
Computers, Quickturn
Design Systems, Exabyte, and NeoMagic. With the
publication of DEVELOPING PRINTED CIRCUIT
ASSEMBLIES and ASIC BASICS: AN INTRODUCTION
TO APPLICATION SPECIFIC INTEGRATED
CIRCUITS, she begins a second career as a technical
writer. Ms. Rhodes lives in San Jose, California, with her
calico cat, Calpurrnia. She invites you to visit her personal
Web site, www.elainerose.com, or write to her at
lannirose@gmail.com.

Paul A. Spalitta

is a Silicon Valley consultant
who works in a variety of roles
from writer to quality
assurance engineer and
business analyst for high-tech
companies such as Cisco Systems, PalmOne, Eastman
Kodak, and Telocity. Mr. Spalitta lives in San Jose,
California with Alice, his wife, and Heidi, their young
daughter. He invites you to write to him at
paul@spalitta.com.

www.ingramcontent.com/pod-product-compliance
Lightning Source LLC
Chambersburg PA
CBHW021908170526
45157CB00005B/2013